BEI GRIN MACHT SICH IHR
WISSEN BEZAHLT

- Wir veröffentlichen Ihre Hausarbeit,
  Bachelor- und Masterarbeit

- Ihr eigenes eBook und Buch -
  weltweit in allen wichtigen Shops

- Verdienen Sie an jedem Verkauf

Jetzt bei www.GRIN.com hochladen
und kostenlos publizieren

Die Auswertung der molekulardynamischen Simulation

Bibliografische Information der Deutschen Nationalbibliothek:

Die Deutsche Nationalbibliothek verzeichnet diese Publikation in der Deutschen Nationalbibliografie; detaillierte bibliografische Daten sind im Internet über http://dnb.d-nb.de abrufbar.

ISBN: 9783346843012
Dieses Buch ist auch als E-Book erhältlich.

Druck und Bindung: Books on Demand GmbH, Norderstedt Germany
Gedruckt auf säurefreiem Papier aus verantwortungsvollen Quellen

Das vorliegende Werk wurde sorgfältig erarbeitet. Dennoch übernehmen Autoren und Verlag für die Richtigkeit von Angaben, Hinweisen, Links und Ratschlägen sowie eventuelle Druckfehler keine Haftung.

Das Buch bei GRIN: https://grin.extdb.e-fellows.net/document/1334264

Content

1   Experimental Approach

All general parameters, which was used for the molecular dynamics simulation of methane are listed in table 1.

Table 1: General parameters for the simulation runs.

Parameter [unit]	Value
Run duration [ps]	500.00
Time steps [fs]	1.00
Number of steps	500000
Number of molecules	1000
Molar mass [g/mol]	16.00
Degree of freedom	3
List cut-off radius [Å]	16.00
Potential cut-off radius [Å]	14.00
ε_{ij} [K]	148.00
σ_{ij} [Å]	3.73

The molecular dynamics simulation of methane was carried out at four different temperatures namely 130 K, 150 K, 160 K and 170 K. For this propose the software *Moscito 4.180* was used. After the run simulation a couple of output-files for the evaluation of simulation runs were obtained. In a next step the density and pressure profile was analyzed and at the end three important files for evaluating the simulation were obtained:

vir.out

fold_dprofile.dat

fold_pressprofile.dat

1.1 Surface tension

For obtaining the surface tension the file *fold_pressprofile.dat* is processed. For this reason the tangential and the normal pressure are plotted as a function of the z-axis at the applied temperatures (Fig. 1- 8). The values for $p_{T,min}$, $p_{T,max}$, $p_{N,min}$ and $p_{N,max}$ are listed in tabular form. The program also applied the Heaviside function in a formulation of *Irving* and *Kirkwood* to get the parameters p_T (tangential pressure) and p_N (normal pressure). After the surface tension was calculated as the area between the pressure profiles for the tangential and the normal pressure, it is displayed in the *vir.out* file.

All results are listed in table 2:

Table 2: Results of the pressure profile analysis.

T [K]	$p_{T,min}$ [MPa]	$p_{T,max}$ [MPa]	$p_{N,min}$ [MPa]	$p_{N,max}$ [MPa]	Surface tension [mN/m]
130	-10.86	0.24	-2.20	-0.15	7.91
150	-4.09	1.64	0.00	2.38	4.28
160	-1.94	1.98	0.82	2.28	2.84
170	-0.21	3.29	1.39	3.61	1.01

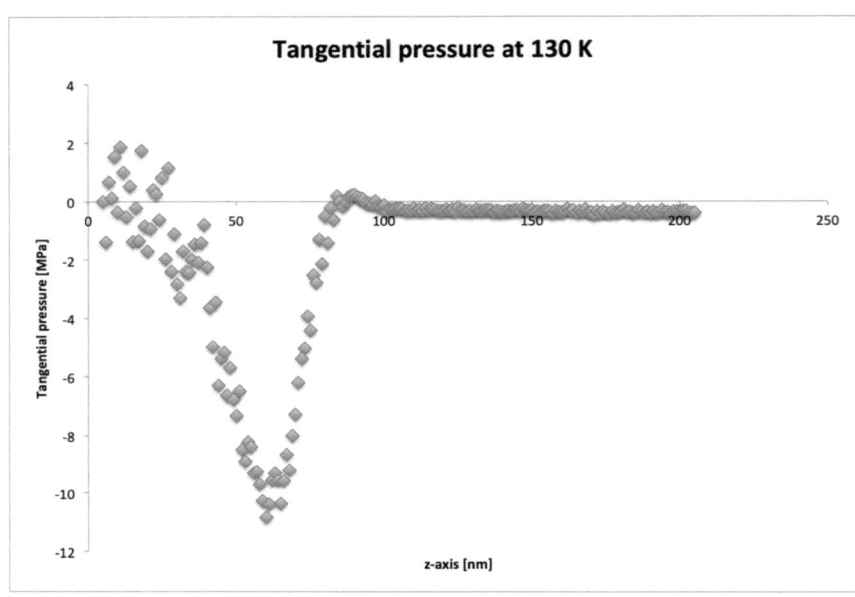

Figure 1: Tangential pressure as a function of the z-axis at 130 K.

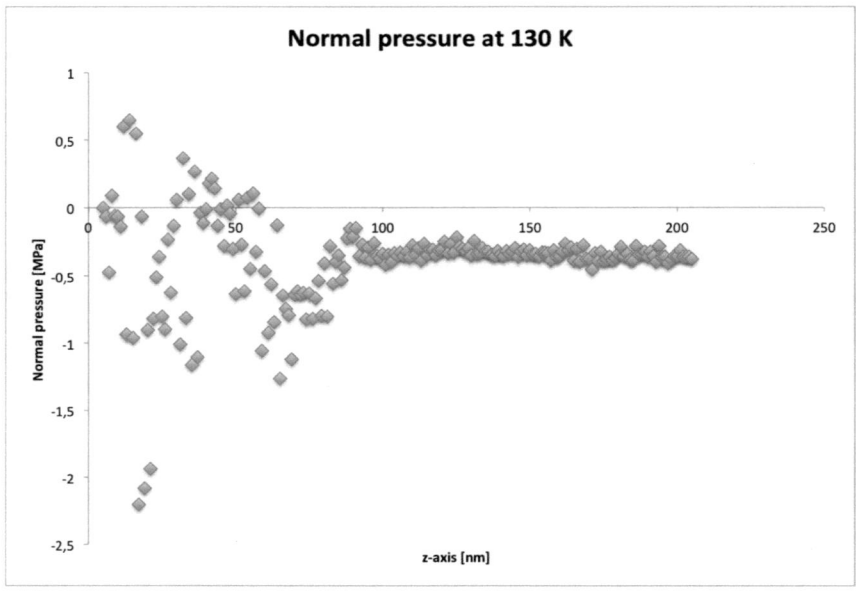

Figure 2: Normal pressure as a function of the z-axis at 130 K.

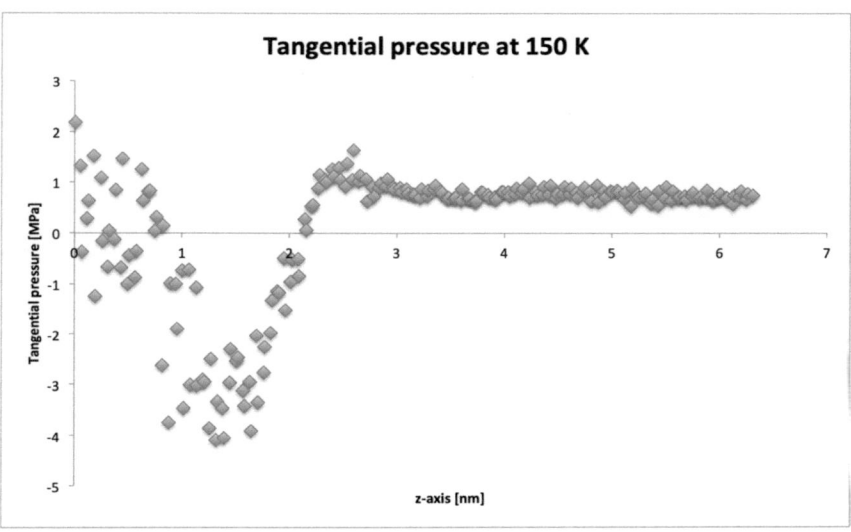

Figure 3: Tangential pressure as a function of the z-axis at 150 K.

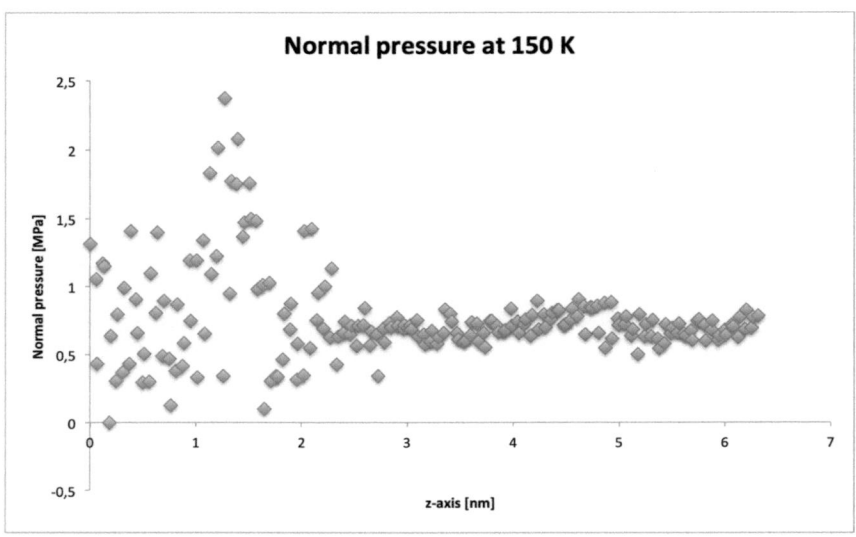

Figure 4: Normal pressure as a function of the z-axis at 150 K.

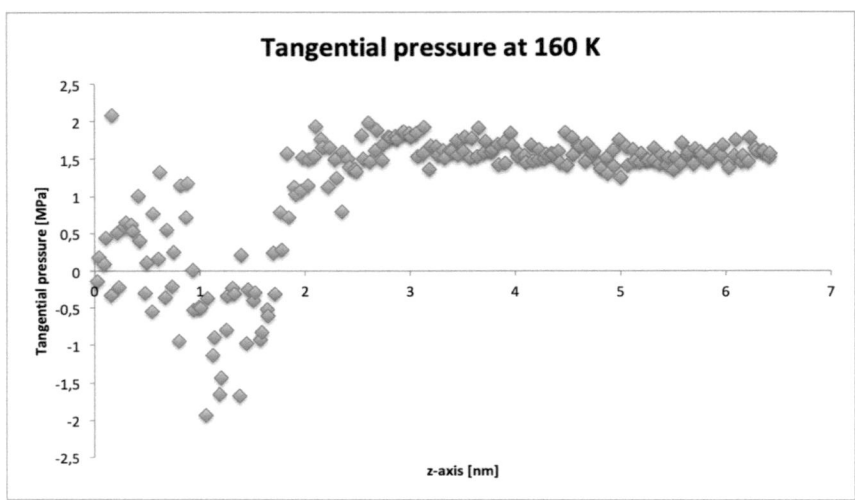

Figure 5: Tangential pressure as a function of the z-axis at 160 K.

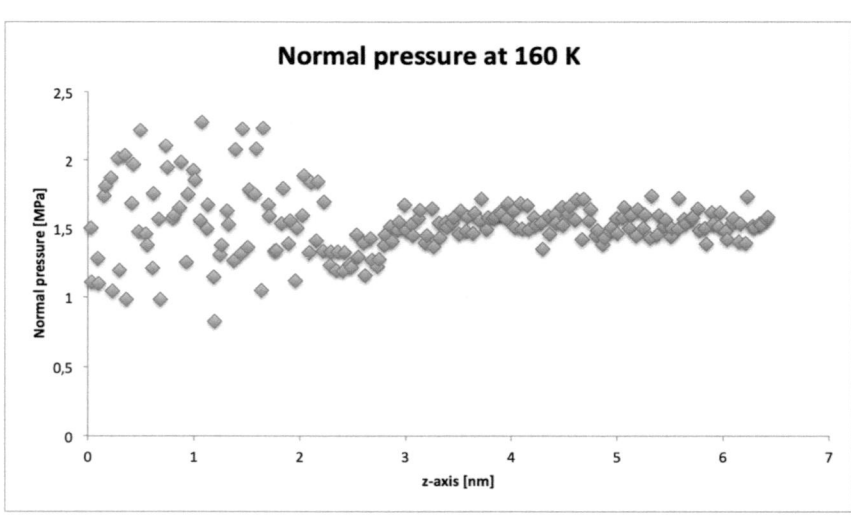

Figure 6: Normal pressure as a function of the z-axis at 160 K.

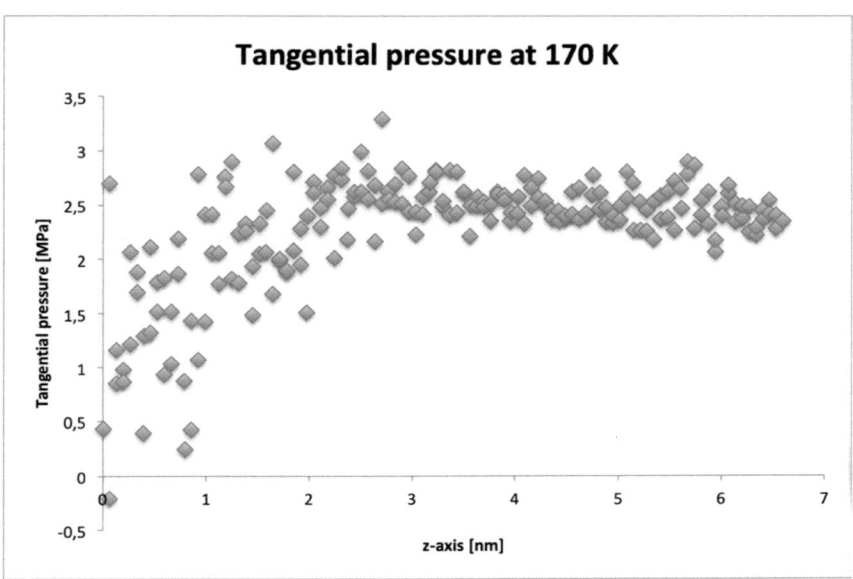

Figure 7: Tangential pressure as a function of the z-axis at 170 K.

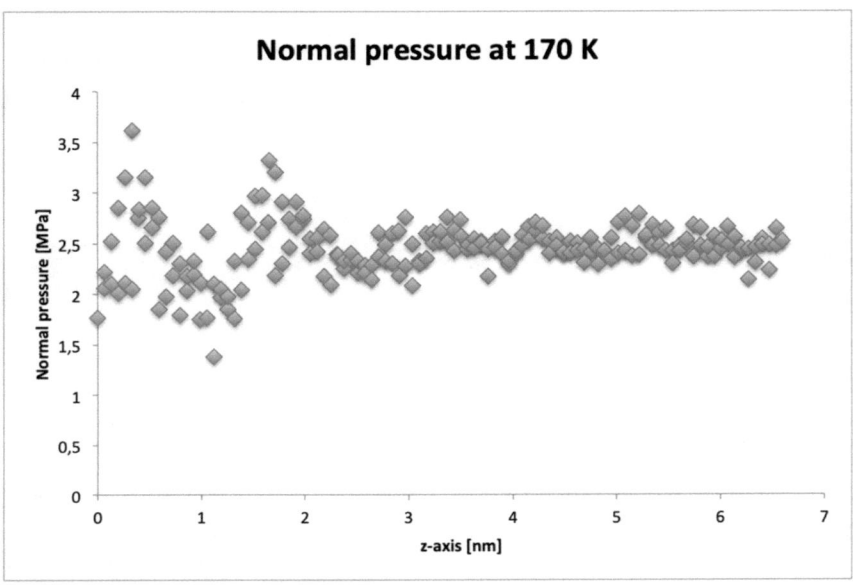

Figure 8: Normal pressure as a function of the z-axis at 170 K.

In a next step, the surface tension is plotted as a function of the temperature. The surface tension is decreasing by increasing temperature:

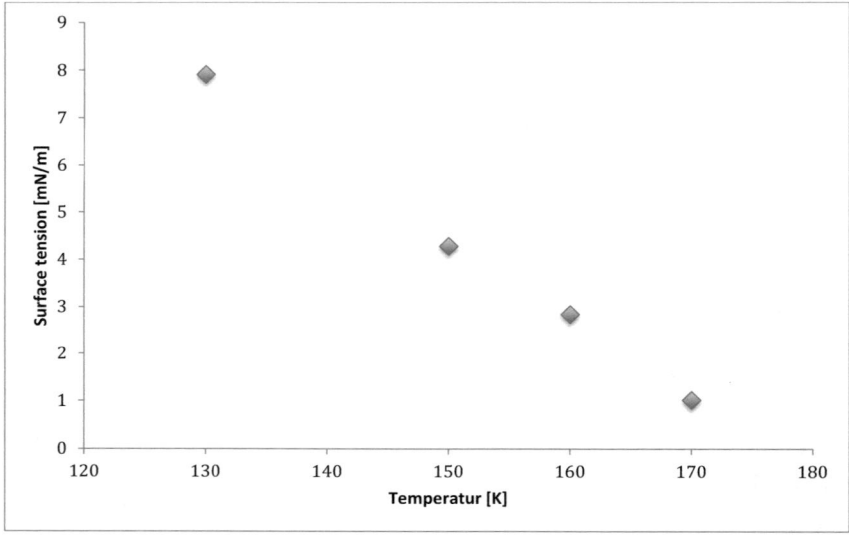

Figure 9: Surface tension as a function of the temperature.

To compare the results to literature data the values are summarized in table 3:

Table 3: Compared surface tension to literature data.

T [K]	Surface tension calc. [mN/m]	Surface tension lit. [mN/m]
130	7.91	9.41[1]
150	4.28	5.75[1]
160	2.84	4.04[1]
170	1.01	2.64[1]

Again, the values are plotted as a function of temperature and the data points are fitted as displayed in fig. 10:

7

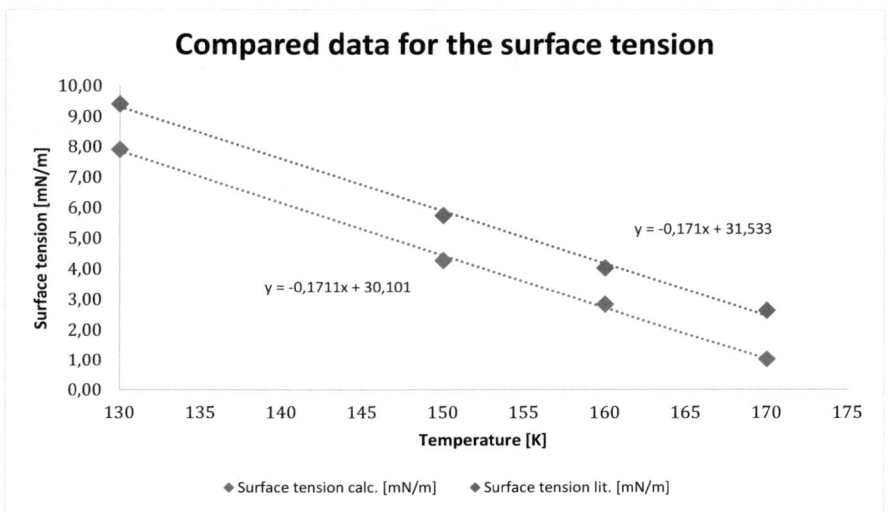

Figure 10: Compared data for the surface tension.

The calculated values for the surface tension at different temperatures are different from the literature data. But they do not differ widely. The result of displaying the slope of the regression line shows that they have the same gradient of the slope (-0.171). For this reason the molecular dynamics simulation in order to get the surface tension at different temperatures seems to be a good approximation to the real values.

1.2 Fitting of the density profile

In order to process the density profile at different temperatures the density which was calculated in the output-file *fold_dprofile.dat* is plotted in function of the z-axis. The plots are fitted with the following formula:

$$p(z) = 0.5(p_l + p_v) - 0.5(p_l - p_v)\tanh\left(\frac{2(z - l)}{d}\right)$$

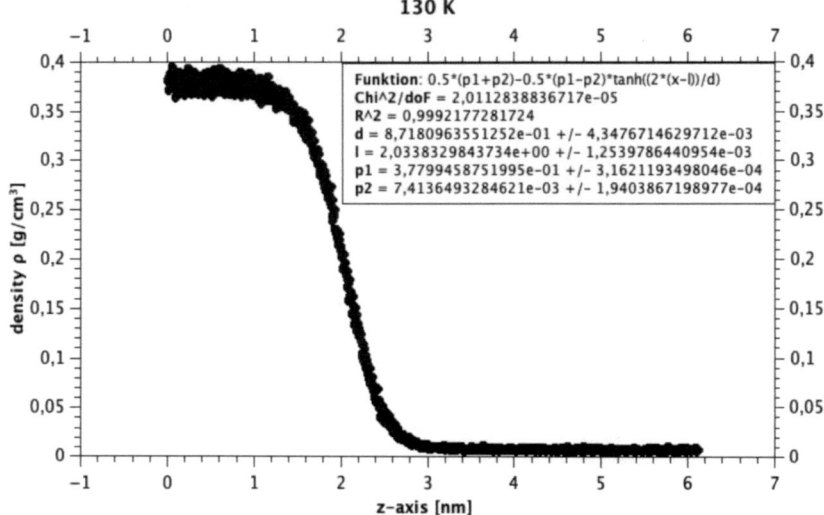

Figure 11: Density as a function of the z-axis with curve fitting at 130 K.

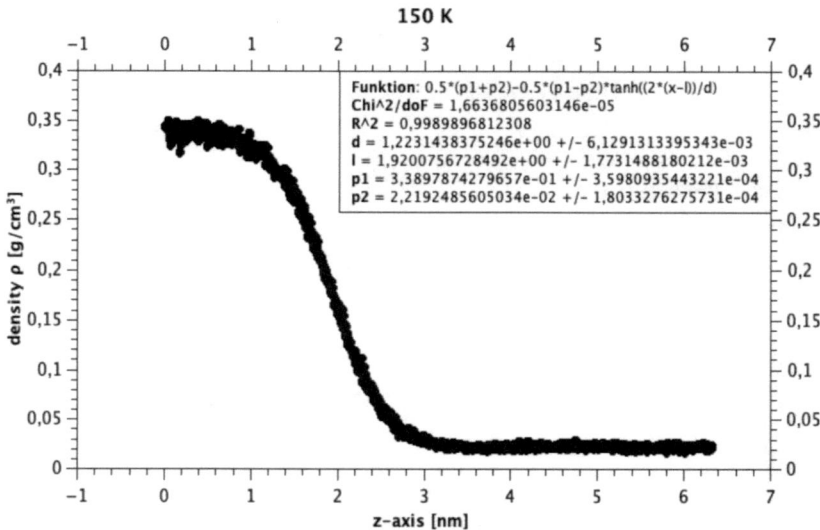

Figure 12: Density as a function of the z-axis with curve fitting at 150 K.

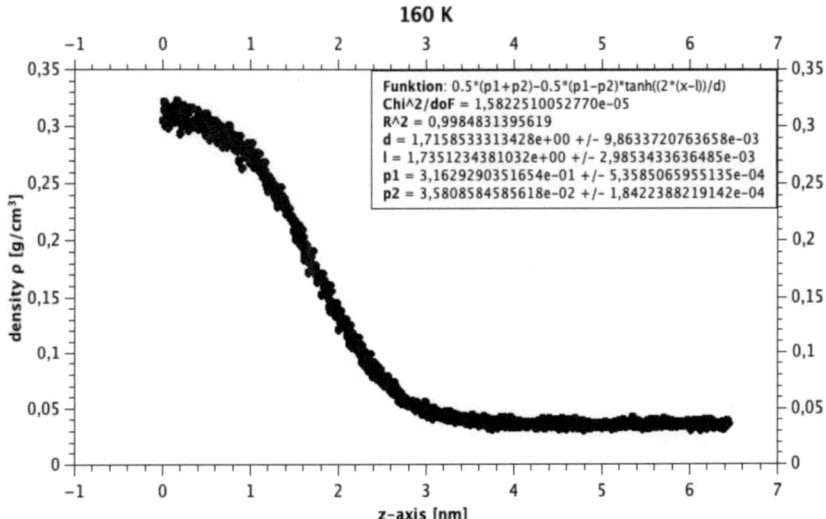

Figure 13: Density as a function of the z-axis with curve fitting at 160 K.

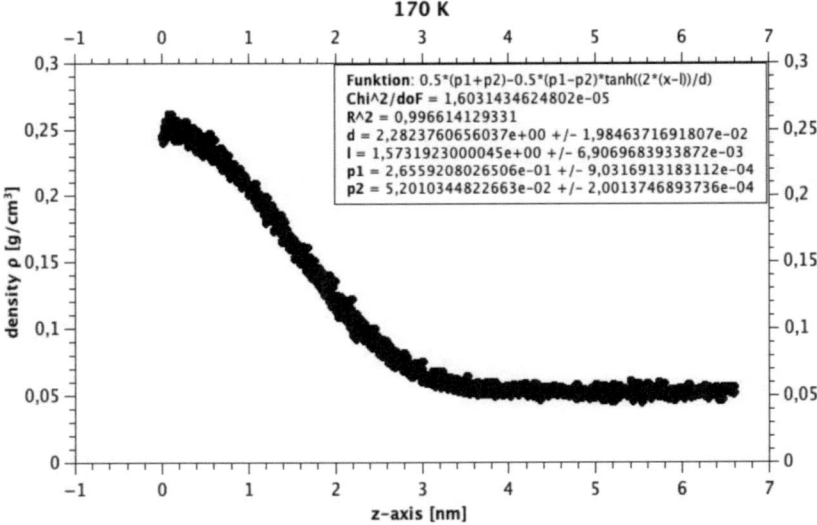

Figure 14: Density as a function of the z-axis with curve fitting at 170 K.

The results of the fitted curves are summarized in table 4:

Table 4: Results of the fitted parameters for the density profile.

Temperatur [K]	ρ_l [g/cm^3]	ρ_v [g/cm^3]	l [nm]	d [nm]
130	0.38	0.01	2.03	0.87
150	0.34	0.02	1.92	1.22
160	0.32	0.04	1.74	1.71
170	0.27	0.05	1.57	2.28

The values show that the density of the liquid phase decreases while the density of the vapor phase increases with increasing temperature. Additionally, the higher the temperature the lower the value for the position of the center of the interface along the z-axis. The values for the thickness of the interface increase according to the temperature.

For presenting the phase diagram of methane, the temperature is plotted as a function of the density of the coexisting phases (fig. 15).

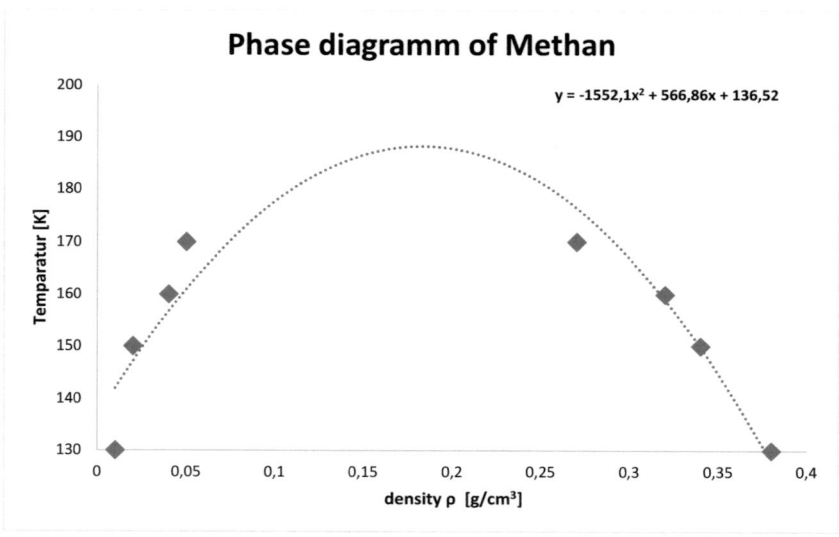

Figure 15: Phase diagram of the coexisting phases of methane.

For determine the critical temperature of methane the phase diagram is fitted with the polynomial degree 2 resulting in the following equation:

$$y = -1552,1x^2 + 566,86x + 136,52$$

11

The equation has an maximum at 188.27 K and 0.183 g/cm^3 so that the value can be assumed as the critical temperature. The value for the critical temperature in literatures is 190,56 K[3]. Compared to that value the calculated value of the critical temperature seems to be in good accordance.

Snapshots

For all simulations, snapshots of the molecules in the coexisting phases were taken (fig. 16 – 19). The higher the temperature, the more molecules are in the vapor phase.

Figure 16: Snapshot of the molecular dynamics simulation of methane at 130 K.

Figure 17: Snapshot of the molecular dynamics simulation of methane at 140 K.

Figure 18: Snapshot of the molecular dynamics simulation of methane at 160 K.

Figure 19: Snapshot of the molecular dynamics simulation of methane at 170 K.

2 Discussion

In experiment the molecular dynamics simulation of vapor-liquid interfaces of methane was carried out. In a first step the pressure profile was processed to get the surface tension. The surface tension of methane decreases with increasing temperature. These results are in good accordance to literature data for the applied temperatures [1][2]. In a next step the density profile was processed to get the phase diagram of methane in order to calculate the critical temperature. The calculated critical temperature is 188.27 K and is also in good accordance to literature data. [3] The small difference to the listed literature data may be explained by the different types of determining. The values obtained as result of the experiment are calculated via a molecular dynamic simulation while the literature data are measured with proper methods. The snapshots of the liquid-vapor interface also confirm the calculated results for the density. The higher the temperature the higher is the density in the vapor phase.

3 References

[1] Baidakov, V.G., Khotienkova, M.N., Andbaeva, V.N., Kaverin, A.M.: Fluid Phase Equil. *301* (**2011**) 67.

[2] Baidakov, V.G., Grishina, K.A.: Fluid Phase Equil. *354* (**2013**) 245.

[3] C. L. Yaws: *Thermophysical properties of chemicals and hydrocarbons*, 1. Auflage, William Andrew Inc., New York, **2008**. ISBN-13: 9780815515968.